Letter and Number Correspondence

Published by Nick Betar

Copyright 2014

Contact: c/o Australia Fair post office, Australia Fair,

Qld 4215, Australia.

I0115668

ISBN 978-0-646-92831-9

Table of contents

1

~~~~

Preface

This book is a researched and attempt at finding the aspects to the whole event that is numerals and numbers. Constructing a reduction of the variance of the cube in correspondence of architecture and technology in application of letters and numbers.

Chapter 1. The Alphabet and numerical terminology

Would you believe that the sum total of alphabetical and numerical information can be present through this very algorithm below in *illustration 1*. The manual layout of this alphabetical/numerical algorithm has with it an exceptional amount of information. For instance, where the master numbers converge on this algorithm, the numbers that make it happen are positive and negative in nature. They are the binary numbers zero and one. They can be looked at in the metaphysical sense as male, the one; and female, the zero, another way would be the phallic symbol, the one; and the circle of life, the zero. To this as translated through the binary numbers, the converging, meaning lining up, as above and below, we have result of this master number through six plus five, equalling eleven. It so happens to be sequenced three time across the algorithm on both sets, above and below on the number line, though not the alphabetical line. F,P and Z are part of the convergence of the master numbers.

Illustration 1.

1 2 3 4 5 6 7 8 9 0 1 2 3 4 5 6 7 8 9 0 1 2 3 4 5 6

a b c d e f g h i j k l m n o p q r s t u v w x y z

0 1 2 3 4 5 6 7 8 9 0 1 2 3 4 5 6 7 8 9 0 1 2 3 4 5

The break-down of the algorithm is started through the knowledge of numbers and letters used in a gridded format, based on the space around us. The space is commonly known in metaphysical knowledge as the 'ether'. The ether is translated to our planetary bodies for example, the Earth, the moon and other planetary bodies in the solar system. This becomes the history of mapping the planets of the solar system, for example Galileo who began this course of action with Telescopic devices using precise measurements in his instrumentation.

We can of course go further back in history and recognise further developments in measuring the planetary bodies and this has been found through many cultures of this type of work. What this indicates is a map, a map that is since it's history of mapping the planetary bodies, gives us, the Zodiac, though before this is considered and is indeed part of this, the binary O and 1, is the bookend explanation to both the zodiac and the colour tone and frequency indwelling within those. It has also the base of those, the colour tone and frequency for a template to whatever structure as I outline further in my book with an explanation of architecture in relation to the letters and numbers. The Zodiac is made of twelve signs, each representing a part of the night sky, as pre-recorded through history, this brings the concluding factor into this.

Chapter 2. Eleven becomes One hundred and eleven

The conclusion of having reached eleven is to realise it's further migration to a higher number, even a vibratory note if a frequency is being measured. Zero, the circle of life gives the charge to the One or the phallic symbol, what resonates here, is Earth bound, where as what resonates in the 'ether' is space bound. As has been mention by Hermetic philosophy, as above, so below, that is the opposites.

In regards to opposites and equals, the quinte-essential problem is determining when a number translates or for that matter transmutes, in the metaphysical it is more determinable, in theory or formula, the same can prevail, at this point it becomes mathematical and maybe complex, so the algorithm encompasses so far, numbers, letters, theses being the symbols, then the link to binary, the phallic symbol the number one, the male and the circle symbol, the circle of life the female onto then the planetary in how numbers and letters were used to indicate measurement of instrumentation as in astronomy and astrology with Galileo and with many cultures around the world measuring and receiving ideas from these numbers and letters. I will exemplify further what I'm reiterating here in the next chapter.

Chapter 3. The idea of 'being' of the numbers

The idea is of the being of numbers and letters made into a word of understanding. Our universe is vast, we cannot assume we are alone in it. Why do we assume sometimes, the vast majority of things that stop thought is a non committal and non communication level on things, a very human trait! But we can break this from time to time, we can experiment, this is human nature and we can have the work involved making us feel all the many traits and ideas but never concluding on an idea, the trait is more of a habit being either negative or positive or neither of the two.

The 'ether' holds the idea, we just have to recognise it and is a creative factor, in fact probably the most creative factor of traits because it's just a word, like an idea. The idea is a number, the number is twenty-six. I'm outlining the obvious, though there appears to be in older texts cultural or otherwise of further letters, the average today represents the truth of our languages, all sharing the average letters. Now, this determines the number on the algorithm that is exemplified to this. Twenty-six represents twenty-six letters, these twenty-six letters are representing the World's, these can be represented best as is heard from Hermes Trismegistos, "The world above and the world below." The world above and the world below have the representation of the Sky or Heavens, the world below is a representation of the Ground or Earth, whatever is closest to the ground is what we participate in, life, the world, what we use to participate with it, are letters and numbers, these are from the ground source. I have shown on the logarithm, the one is for-most representing the master numbers, being that of eleven twenty and lastly thirty three. Binary represents space or void. What makes for the ground source is the number one, or male for the source.

What makes for the Sky or Heavens, is the number that constitutes a factor of divisible though in the algorithm, the invisible, zero and zero only equals, the circle of life. It has to be made into a number twenty-six! The reason is binary, the application of space. Twenty-six becomes two-hundred and sixty. Semantics on this idea can be applicable anywhere between twenty-six and two-hundred and sixty, to cut this short, you can now add a zero, a circle of life to each concluding sphere of space, much like an onion. This explains the grid of space and environmental factors and so the grid will contain other new logic, understandings and principles. Varying consciousnesses that I call 'illusory' being illusion operate in the dynamics of mankind's contributions to the physical realm, for example, the impact and imagery from television and the sound produced from a radio, any varying logarithmic patterns operate at subtle frequency giving these phenomenon and all operate at any given time in these realms. The extent of these phenomenon and at which and what goes on each generation, is the changing patterns within each grid learning and their structures of uniqueness within each structure, there is the question of the cube in relation to the grid.

Chapter 4. The cube

The grid has the three dimensional grid cube attached to each grid square, for where ever there is a square, there will be a cube. The cube and not unlike ourselves of that individual space that both the cube and ourselves may occupy, has other grid cubes that are either larger or smaller, yet no less significant, grid cubes of consciousness, are for the grid cube that we exist within at whatever level, height breadth and depth, they are always around us at any one time.

As in algebraic equations, the set group of numbers is apart from the primary of the group, example, $7 - 6 (+1) = 2$ yet if this is changed to $7 - 6 (-1) = 0$; What accounts for the difference is 2 and again this $7 - 6 (+-1) = 0.5$. Overall, the same goes for multiples of a number, for instance, a 0.5 or 1 added to any number in a set group or individual number; another example, we have $7 - 6 + (3 \times -1) = -2$, although change this to $7 - 6 + (---1) = 0$, just as the opposite $7 - 6 + (+++1) = 2$. If now the character is individualised and so bracketed as $7 - 6 + (-)+(-)+(-)1$ as individualised separate groups for -1, it is then -2. The same in positive is $7 - 6 + (+)+(+)+(+)1 = 2$ This demonstrates a - 1 or +1 are one and the same from the perspective of the cube and of space in performance with numbers and letters of the algorithm and the Illustration below, the individual cube is both negative and positive, it is also as demonstrated with the algebra equations, indivisible to multiple performance of cube interaction. Energy manifestations would be naturally random and so a different reflection but the same grid.

Grids represent language and language represents our reflection of where we are represented in the space of language, the dimension of it. The world holds the only answer to our organic structure.

Chapter 5. Mapping through to architecture

Simplicity is the apparent determinant including that earliest of mapping of any project that is undertaken as for example, a boundary line on a field to the latest of satellite technology taking a picture from space. The same goes for general planning because history is integrative our whether we do something on a small or large scale. Through history, mankind has endeavoured and successfully undertaken these opportunities. Earlier on in history, theses have started with simple prerequisites, be it based on tools for a base to map and build a house and integrative features of practicality.

Here the number system has worked from earlier times using mathematics, geometry and set squares, right to the present day in delivery of functional modes of working , these however were not always as paramount as they are today in the use mapping and architecture using computer and technological processes of design and architecture. Programmes such as 'Cad' are in use for such requirements of making and designing, the same is related to the technology, for example aircraft, though, this being more recently.

Flight and technology is and has integrated throughout our history alongside the developments of mapping and architecture. The early pioneers of flight brought us into a recognised history of flight. In ancient history, the myths of Indian scripts of texts tell that the working of flight was prevalent back many thousands of years ago with a vehicle, the scripts tell us of their Vimana's, being saucer shaped. Because our history is mixed with accounts whether historically mythical or otherwise, we are never the less dependent on this to further development in areas of expertise of thought and practice.

Numbers have played a pivotal role in historical workings, this development has shown another feature of the number on the top line of the algorithm. As illustrated;

Illustration 2.

1 2 3 4 5 6 7 8 9 0 1 2 3 4 5 6 7 8 9 0 1 2 3 4 5 6

I've outlined just the top line to show the correspondence with our present history.

The history and future, if we could say are one and the same, then where would the catalogue of the Akashic records be housed and what would they look like and like architecture, what would the structure look like? For the formula of mathematics or the length and structure of written words.

This shows the break down description of what might be Earth. void and pressure=darkness and Air is also void with no pressure=light. Therefore for example the letter 'A' holds information, the same would go for the number '1', though what can be vice versa is the pressure or no pressure, for example, an algorithm generated piece of information code that is put into a computer is going to have more pressure with more numbers or letters or both then say with fewer or just one number or letter, this alternative description is to do with 'space' on a computers drive although this can be looked at as a type of technological microcosm of the greater space outside of that realm. So information that can be the same or one character or

6

unit or units, again for example A is A as a unit through to Z or in algebraic terms, A is one through to zero as a group, depending on how you approach the usage of the unit or character, the former could also be multiplies of the letter or number as another example, algebra produces sub-sets with mixes of numbers and letters and this could be of an algorithm.

Taking into account the logarithm, the unique structure of it is prevalent to equal and odd numbering taking into account letters. It's usage in pyramid design and structure will have a zero and a one in the building of the sides of the structure. There is a triune usage being three points in architecture. In this matter, mankind has discovered knowledge, placing it into engineering in letter and number, to give the practical exemplification of universal structure and which is the triune, the three points in nature as the basis for structure being solid and symbolic.

Working from below the ground to the surface and then to air, there is interpretation to each of these areas. From the physical and the metaphysical point of view, below is pressure, it is dense under the ground from rock structure, it is mounting pressure the further you go down. The surface, our landscape is unlike the ground below, though similar pressure or air pervades here and air has different pressures, Earth, Air, Fire and Water make up the basis of the compositions of these, with different chemical and composition both to the rock structures and the air and gas pressures, with all of these having a gauged number and periodic letter, the algorithm we have technology utilizing Earth, the breakdown of materials into their components, then go into building structure, using construction materials which is for architecture, then mapping is the structure of the inner architecture through mind and intellect, the inner architecture description, location geographically and utilizing the algorithm number and letter as formerly described both in scaling and drawing the print plans, the third is air, it represents technologically, flight and the structure is aerodynamic and whilst it arrangements fits into aerodynamic shapes and materials, the structure is born.

The Quabbalistic term for the usage for name used for God has the expression of a zero and a one expressed. This is a good example of for instance in the Akashic records, a sub-conscious reasoning for what God is and we have structure for the records. To my mind, the structure, indeed the architecture of the Akashic record, our spiritual depository is none other than the zero and the one being a hole in the structure of space. This could be termed, a portal or an opening and almost a metaphysical meaning with a biological overtone, again as mention the zero is female, for 'vagina' and the one is 'penis', the knowledge could be termed as this. Whatever this transmutes into, could be just called knowledge and to go further into terms about this, is back to the present, the present represents Earth matter and out of minerals from getting the out of the Earth, mined, drawn up in the plans and made to be the completed product from Earth to Air and again drawing on the factor of Fire and Water in the displacement of the working materials, the rest is mapping it out.

Culture is predominantly elucidated from its history, many new and not so new inventions in the field of architecture, written workable plans for the design and manufacture of Earth, Air, Fire and Water in the product sense is conclusive engineering and found at ground level by this I mean that we have a formula which uses algorithms or hints at that to use the terminology and can be linked to Phi or (golden mean ratio) 0.1618 and into infinity from the numbers from there. When the circle is squared, six represents natures formation. as with *illustration* 3 of the golden mean below:

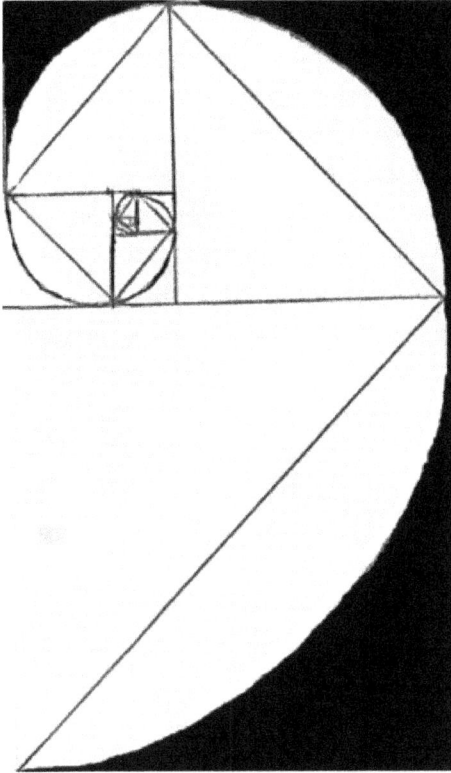

Illustration 3.

It can be used as a type of measure in architecture gauging it is neither one or zero, it could be all the numbers in connection with the golden mean, unwinding and winding because of transfer of information this on a biological scale. The nautilus shell demonstrates that to build the shell took this factor into account as the sea creature grew. Like the zero and like the one, it is neither negative or positive, they are interchangeable yet hold to a set of principles and laws. In theory, like a strand of DNA, that winds and is a portal, it is an opening to the Akashic awareness within.

Illustration of geometries of the zero and the one. What they constitute. How they function in terms of geographical and geometric design in nature.

Illustration given below of the set of separate but inter-related geometries to structure and sub structure on microcosmic and macrocosmic levels in relation to atom the fields of algorithmic inter-relation and interpretation. Assuming there is no wall around the structure that blends within this, the cell is never a cell without a structure that is understandable to human logic and thinking. There is a membrane that surrounds this and there is an energy field within the nucleus. They confer as to the following with *illustration 4..4A.* Shows the perpendicular lines of structural core. *4B.* Reveals as I've outlined, the inverse and obverse of the pyramid structure with pyramid tips touching each other or the point within the circle structure. *4C.* If the pyramid is drawn within the structure, then it shows the more adapted structure of the Tetrahedron. *4D* The structural core then has around it, the tetrahedron structure, which in its structure also shows the seal of Solomon. That which is below is also above. It demonstrates how the circle helps to manifest this above and below from a side view of the pyramids with tips touching the point.

Illustration 4....4A,4B,4C,4D

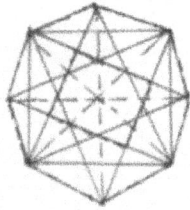

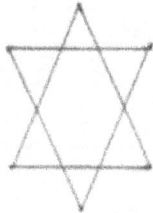

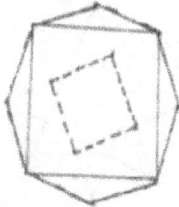

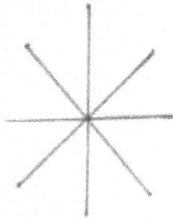

4A,4B,4C,4D,

4E,4F are further exemplifying of the same showing different regions of use in geometry

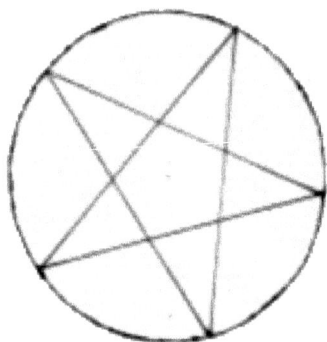

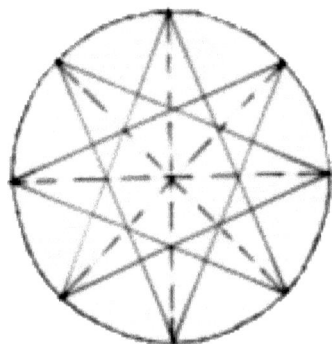

4E

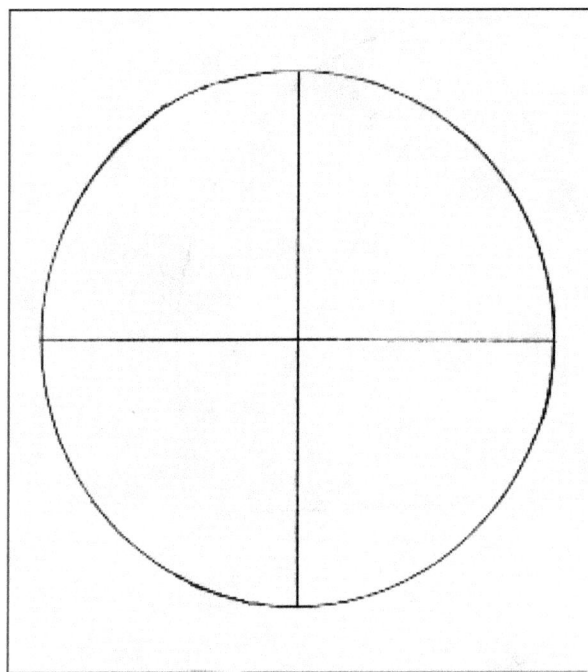

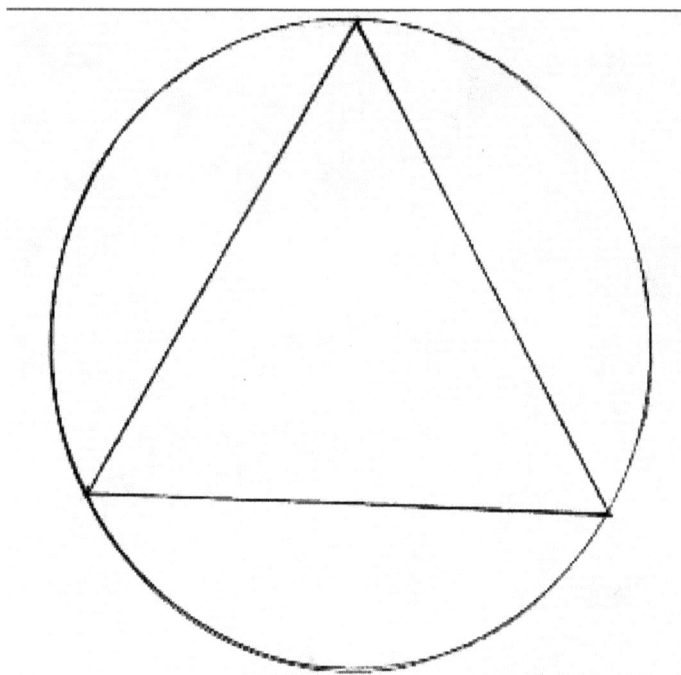

4F

14

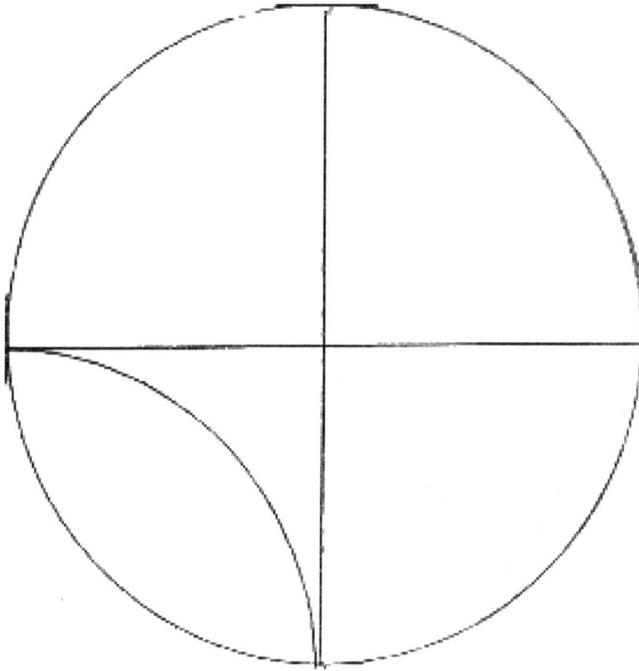

4G

This illustration is comparative of the flower integration with a semi circle added to the squaring of the circle in one quarter panel of the illustration of 4G. It is indicative of the continuance of structure both in nature and architecture, which of course parallels in mapping, geometry, (geology i.e.: geomancy) and technology, these include the application of them. The quarter image if turned on its point, pirouette, the image assumes the typical squaring of the circle image.

There is an interesting doubling effect with the number sequence that is prevalent in the numerical/alphabetical distribution of the algorithm. For instance for the number of 5's on the top line of the algorithm is 15, the 6's on the bottom, make 18. 1+5 is 6 and 1+8 is 9. This would have it symbolism then in the balance of ying and yang, interestingly enough. Here are some equations linked to this algorithm. Here's how some of them subtract out as an example;

Illustration/description 5.

12/12   12-12 = 0

15

12/34   34-12=22 breaks down to 4

12/56   56-12=44 breaks down to 8

12/78   78-12=66 breaks down to 12 then 3

12/9   9 - 12=-3

incidentally -3 + 3= 0

56/12 Twelve representing the Zodiac is 4.666666667.

In keeping with the algorithm layout, the five and six get its zodiac division in the sequence, numbers four, five and six, the six being repeating eight times. So the sympathetic resonance of numbers, shows this as eight after the seven, this demonstrates the close and seamless proximity of the planetary zodiac, in fact one and two is three beginning the number sequentially, the six relates to the nine being what seven and eight are added being a six, just upside down nine. If seamless, then this illustrates that five also relates to the female and the binary circle which is female and the six and the one for the male.

Chapter 6. Twelve represents the one in twelve

(3, 4, 5, 6, 7, 8) works from the principle of multiplying of numbers. Between the breakdown of the four and the twelve in the sequence in illustration 6. we have the number twenty-four, this fits in with twenty-four hours in the day. Base for square using this table;

1 x 2 = 2   2 x 0 = 2   2 + 0 = 2

3 x 4 = 12   1 x 2 = 2   1 + 2 = 3

5 x 6 = 30   3 x 0 =3   3 + 0 = 3

6 x 4 = 24   2 x 4 = 8   2 + 4 = 6 Note; this is in reference to 24 hours day use.

7 x 8 =56   5 x 6 = 30 30 + 0 = 30 = 3

9 x 0 = 0 9 x 0 = 9 9 + 0 = 9   = 9

If you add all the end numbers, inclusive of the noted reference to 24 hour time, the resultant numbers being, 2 + 3 + 3 + 6 +3 + 9 = 26 the exact same number as the alphabet which the algorithm is based on! And minus the 24 hour reference which is the 6, it is 20 though inclusive of the 6 it is showing it is synonymous with the end result of both the alphabetical and numeracy characters.

With the illustrated example following, I used the synchronicity of numbers to blend in where;

$2 \times 2 = 4$ $4 \times 0 = 4$ $4 + 0 = 4$

$8 \times 8 = 64$ $64 \times 0 = 64$ $6 + 4 = 10 = 1$

$10 \times 0 = 10$ $1 \times 0 = 1$ $10 + 0 = 10 = 1$

With this you get to the remaining $4 + 1 + 1 = 6$. The synchronicity follows through into the 24 hour time.

Chapter 7. Key coding numeracy

The way in which the universe may operate according to the algorithm of alphabetic and numerical numbers, is to stretch out, much like a continuation of the algorithm, much like a chain of zero's. If space has area time has, width, height and breadth, then you could add a zero. Twenty-six would be, two-hundred-and sixty. So with the *illustration* 6;

2 6

2 6 0

2 6 0 0

2 6 0 0 0 add infinitum.

This would show expanse within a universal spacious field. This would inevitably be space using the same geometric layout as before, that which was described with the chapter on 'the cube' function. It could be likened to a double helix pattern in the DNA, like a strand molecule!

This molecule has hypothetically/metaphysically speaking a memory pattern in it as well. If this being the case, then it should arrive at it hypothetically speaking, more evolved formulation, thus the link in the chain. 26 may just as well be any variant of numbers, say 46. this could follow on with a chain as with our alphabet. The hypothesis being that if there was a more advance civilisation further on the time line, then this would be a higher variant of letter number construct. So to *illustration/description 7.*

4 6

4 6 0

4 6 0 0

4 6 0 0 0 ad infinitum again, completely random, though in context for this hypothesis.

The zero's, I'm surmising to be the catalyst or holding of information, much the like the embryonic stages of a cell before it splits and becomes one or more. The base is the number and could be any number as example 26 our alphabet/ numerology algorithm, depending on evolutionary growth, The following zero's holds the potential to the surrounding universe from a metaphysical or Gnostic point of view.

A life cell at it earliest stages of growth, could be a civilisation/construct of 1 0 to the power of 100 as an example through to 100 to the power of 1000, as another. Please remember this is hypothetical.

When metaphysical activity crosses over into other realms of science and literature, hypothetical interpretation can be used to metaphysically calculate an inanimate object with an animate, physical one.

Chapter 8. Twenty-six in algorithm

Twenty-six represents the alphabet and it's algorithm in convergence with the numeral, or numbers. I've covered the interactive idea that the master numbers have a relevant place in the distribution of coding, that of one and one or a circle above or below another circle. This could be represented as two circles, one above the other, though in a vertical presentation, the eight too, has a close resemblance to the DNA helix. *Illustration 8A and another illustrative comparison of this shown in 8B with the binary comparisons 0 and 1.*

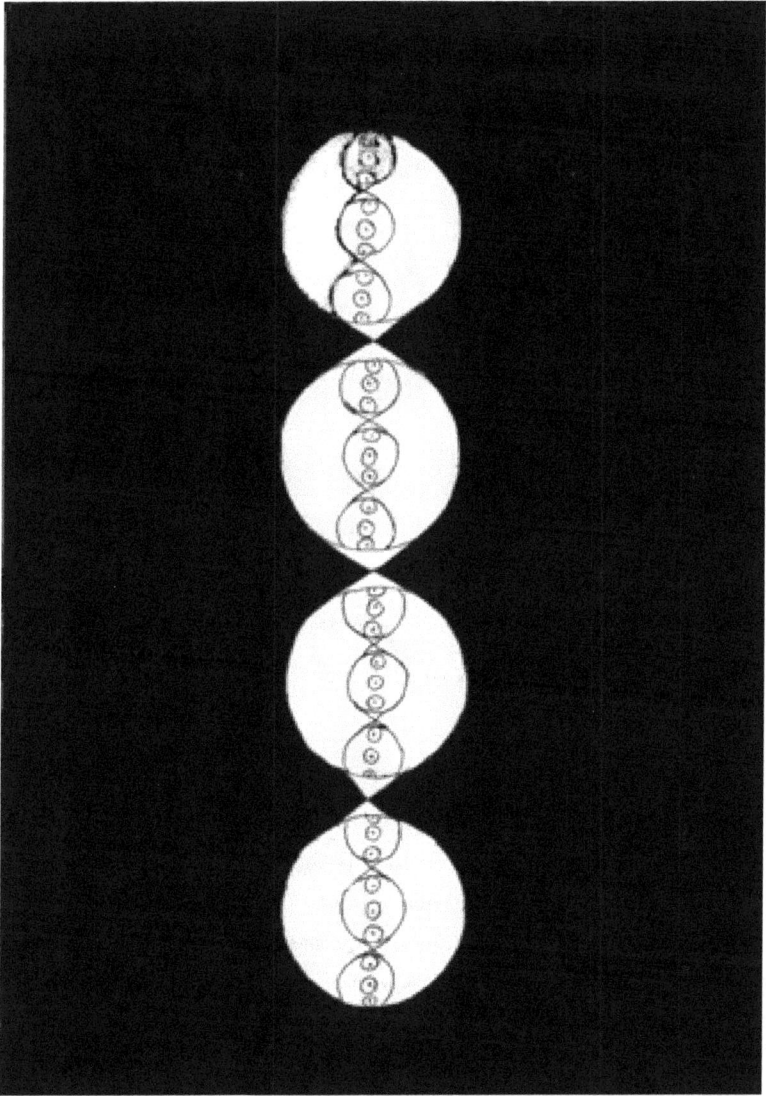

Illustration 8A

19

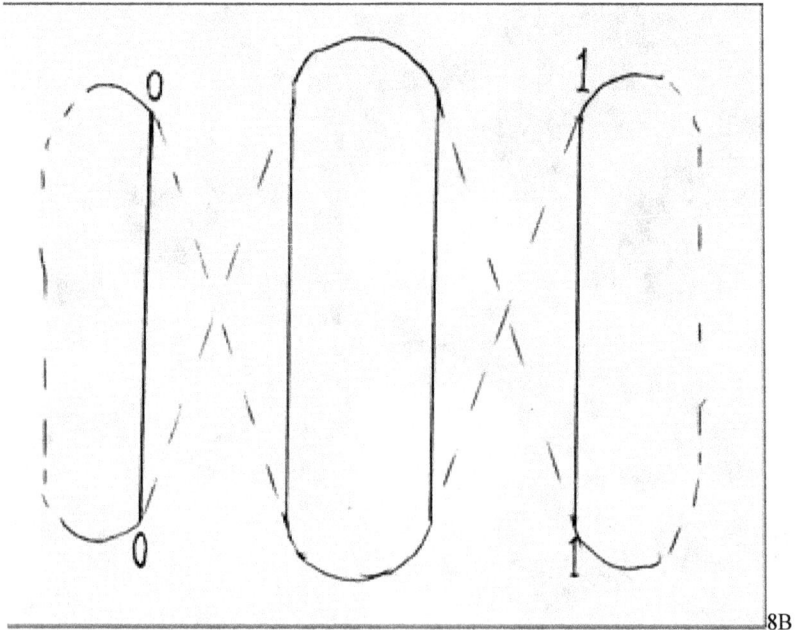

Illustration 8B

0 and 1, notice there is a DNA helix associated with the form. Each link or strand helix, can represent an evolution. This can be configured into the key coding numeracy into the many evolving aspects. Illustration 8A illustrates this in that the DNA helix when looked at can represent the microcosm, that within that strand lies another of the evolvement of that strand and that the macrocosm represent the physical completion of that strand in the microcosm. As above, so below. The DNA connects to the numeracy of the numeral.

The idea also two and six could be in another representation, a parallel to a grid reference and taken into account with the illustration of this in the 'key coding' paragraph, whereas example; civilisation/construct of 10 to the power of 100 as an example through to 100 to the power of 1000, with as many zero's, one's and any of the numbers in between can be regarded and as another by in large, the difference in a civilisation may take on a uniformity as in the case of randomness or un-randomness of numbers as contributes to it civilisation and any wonder, that architecture has a distribution of shape and conformity as well.

Back to the grid structure and it has numerous boxes or cubes, each with their own reality. When I mention, 'their own reality'. I mean it as a paradigm. That isn't to say the paradigm doesn't have other numerous bodies or other paradigms interrelated with it, in the cube of space itself. It has been explained that the universe is crisscrossed with energy currents of different energies, this may have something to do with this phenomenon. As the illustration shows, the squaring of the circle is reminiscent of the factors pertaining to the cube and the pattern of it as a functional form, even an entity. These would be symbolic of the resonance of the patterns, usually in the most basic structural form as is the atom.

The link with universal principle colluding with a path steeped in accumulated and intercepted function, is an apparent manifestation of will. Through this will, the result ensues.

Politics can be the initial spin offs of resultant discovery of fields of the architecture, (the method used for anything from mathematical calculation to building the Sistine Chapel, mapping (which is the consistency of following a path, where we have been, whether marking off the peg for Sistine Chapel to the stone markers of trekkers following where they may have left off in an underground cave.) Flight , (being the function of the idea and reality of flight such as flight throughout our time, mythological Vimana's to technological flight.)and technology, (the technology that has been present throughout the ages, mythological and the consequence of technology in the present time from the introduction of the Industrial age, to the invention from everything from computers to aeroplanes.)

The construct of an atom could be the link between crystalline and biological (animal species) will inevitably be linked to the closest representing structure, a dodecahedron shape of the universe. Thus shaped due to the subtleties of the planetary bodies as opposed to those that dwell on them, the lattice work and nebulae of the universe itself. It is more like a spore, this dodecahedron, a free floating crystal, biological spore. But how could this be? It is a supposed truth, one based on theory rather than true evidence, that if the universe can evolve within the spore (organic) like dodecahedron, then it may not be a stretch to believe otherwise of the idea of a spore (organic) atom. This crystalline growth of the universe growing outward from within and without outward, far are not both crystalline and biological forces at work on air earth water and fire elements? The lines of force in the universe which unlike wolf's law on the explanation of loading of bone the Human or Animal will adapt to the pressures of the bone loading, the biological evidence of stressors on the growth of bone or changing of crystalline nature of it within the body of Humans and Animals and could in this case be true for the same lines and energy of force found within the universe and effect structural growths in the outer macrocosm rather than the inner microcosm. Found within this is the apparent theories of our small universe found within the structure of the universe itself though on a smaller scale. I bring this up with the value of number in Key coding numeracy Chapter7. We are living in a crystal which could have the characteristics of organic and inorganic structure.

*Illustration 9.*of a dodecahedron. The concentric pattern in it is followed by a use of a further delineation and exemplification of cube dimension and space, it also fits in with the illustration 4F and 4 G with squaring of the circle and with the .more so however with 4F in that the triangle has the angle or crystal form, indeed in nature nothing is truly square, when looked at from the microcosm, the macrocosm can show the manifestation of angle, though this is alternating at the changes in these perceptions, as indeed the viewing from a microscope of these things can also testify to.

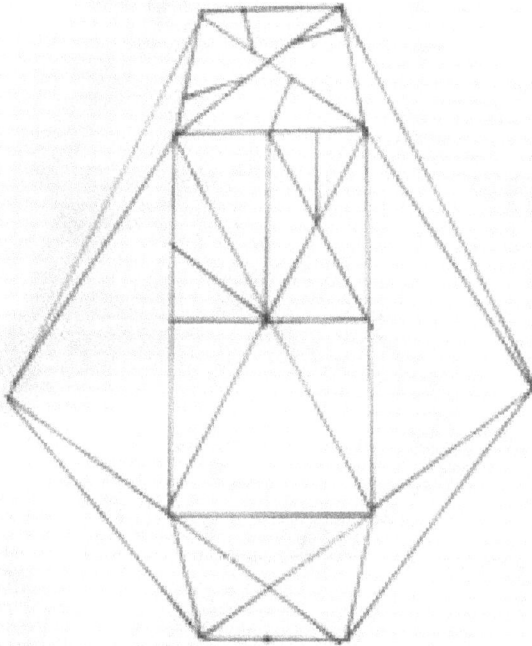

Illustration 9 Dodecahedron.

Technology is based on nature, the nature that is related to God. The God factor is found in natures fauna and flora and is related to the oceans and all that is in them. They are related to finer higher tones of the planet where humanity abides, the lower tones are in the rocks and associated metals in the Earth's crust. Up until it becomes recycled back to un-associated tone with that which is higher and on the surface of the Earth and found above in lower in degrees in between and throughout associated. The terminology, that which is above is also below as said by. Hermes Trismegistos.

The higher vibrations, mirror the lower vibrations. The equal and opposite reaction that drives the planet we live on and her surrounding cosmos. There is an apparent dual purpose in the foundations of the Earth, that must mirror all that surrounds the Earth itself. Many geomantic points on the Earth are in relationship with the numeral/alphabetic algorithm as is the cube of space dimension. A display of this could be described as the squaring of a circle is shown in the next illustration. As in Illustration 10.

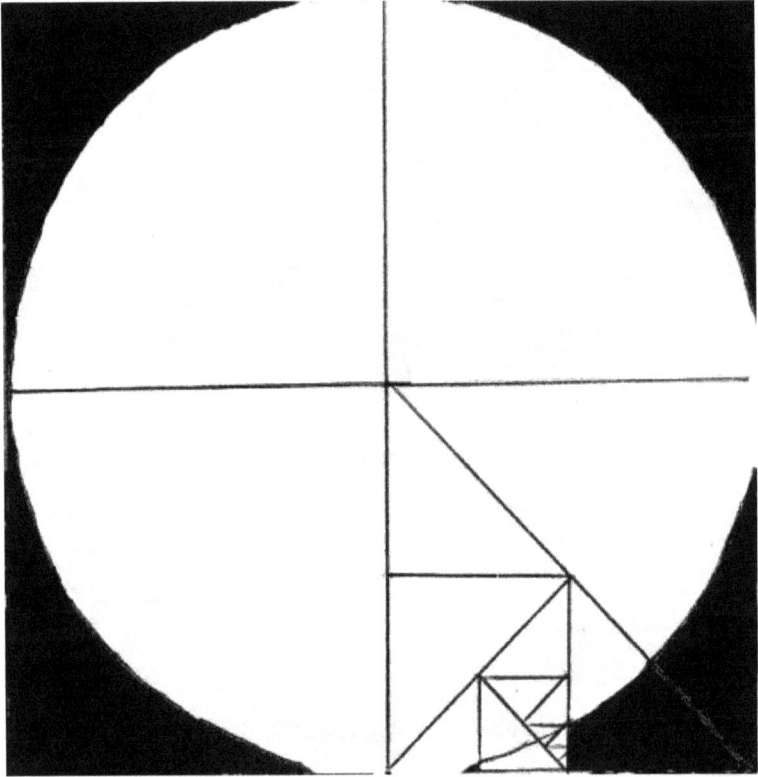

Illustration 10 A field of further delineation into another cube and space

Once a field is worked out, mapped or projected onto a diagram, another set of repetitive structures of the same semblance though in a different space will come up., such as what is found in the above illustration. Not only a circle, now that it has been graphed in diagram, we'll have plotted into it, the squaring, suffice the squaring of the circle and the triangle. This method help with the next dimension of field, that which is off to the right of the *illustration.10.*

In the next description, the display is brought forward, so that dimension is better seen and understood. *Illustration 11.*

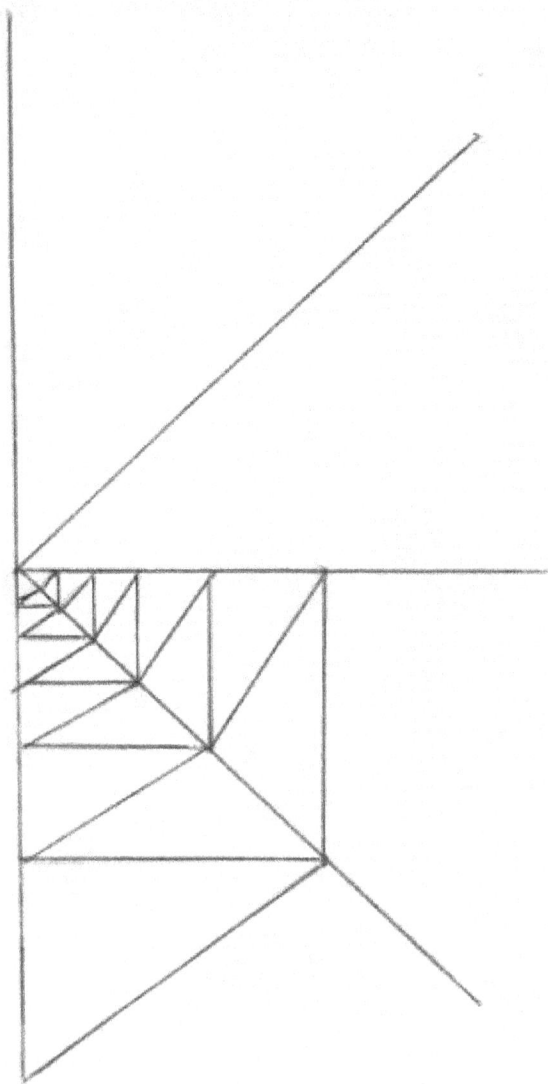

Illustration 11A

26

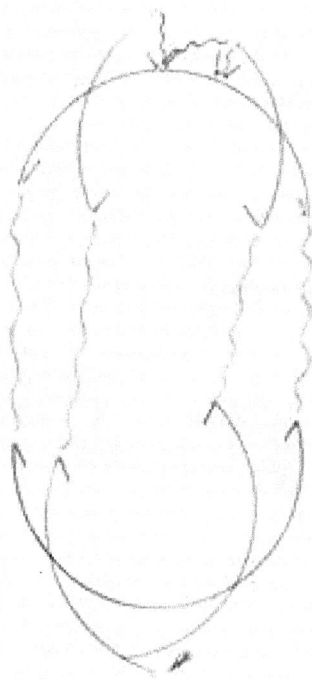

11B

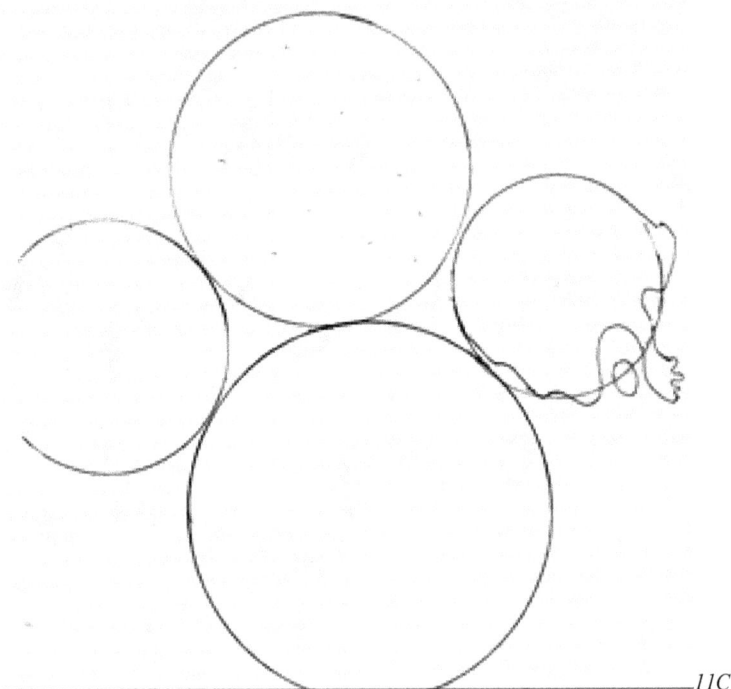

_11C_

Illustration 11C shows that the atom could be mapped as on the surface layer with different reactive traits both positive and negative and also neutral, there is a disassociation with the far right circle (atom), which represent the element of 'form' or conformity of change associated with the atom itself (in theory). It is a randomness of shape and function over-all, for a solid and is the dissociative element of form, over-lapping the same form, which in theory creates a new element in dimension. The _illustration of 11A, 11B and 11C_ shows with the top circle dotted points, these could be the behaviour exhibited by atoms as of the microcosm which could tie in with illustrations 8A and 8B explaining microcosm and macrocosm behaviour and has a further extendable element to the description of the further delineation of the cube of space again using the squaring of the circle. It can be seen here, in the use as in illustration 4, _4A_. Shows the perpendicular lines of structural core, now shown in half value to the squaring of the circle, which is derived from the cube and therefore the square is a singular reflection of the cube, the circle of the globe and the triangle of the pyramid, all inter-related to each other, finding all their values in numeral and alphabet terms.

28

Chapter 9. The spectrum and tone

The colour spectrum of space and the cube is demonstrated primarily by the separation of the primary colours of the spectrum, for example red, blue, yellow. White and black fit into a differing correspondence to the order of these primary colour that in turn can separate further into colouration through degrees of manipulation, such as purple green, crimson brown and so-on. It is dark matter that fits into the mystery with light matter, just in dualistic correspondence both with different applications on the dualistic scheme of thing.

Finally 'tone' has a capacity to work with the correspondence of colours, sometimes keeping these explanations short helps. The tone and colour correspondence has the application of which there are at least three, 'static' colour or tone in a neutral position and/or moving energy will produce a differing result. The dark matter is such because of depth of darker correspondences to the light energy. Some depths are greater than others and create in the deeper veil of the universal fabric, a rift or overflow of energy held together by a sympathetic factor of universal abundance of such matter that it will because of space allocation tone and colour will correspond to that 'frequency'.

Chapter 10 Space bodies and the Mandela

The chapter here I've added after further recollection of researched material. The early period nature religions had ideas and furtherance of ideas regarding the space bodies, the planetary material and why it was established. The universe, the zodiac and planetary influences, indeed anything regarding the planets could be attributed to these early religions and civilizations. The Gnostics explained that the planetary bodies were the work of entity spirit energies, naturally migrating to these regions. As the universe expands and contracts and has always been this way. After created Mandela's express a subconscious delivery of visual information, opinion can widely vary on the nature of Mandela's but it would appear that they centre around the deep religious, esoteric areas of information, explanation of Archon, Archaic deriving can be argued. The deriving of a space-body has to be explained. If an energy from a completely different locality in the universe, travailing from another galaxy, effecting another galaxy in the process, then it is no question of it being linked. How can that be explained now with the region we call the here and now. It is a mystery, though the energy which put itself there in the first place, not being fully understood and for that matter not

29

really taken any interest of at all, place itself there if it's in or on two regions at the same time. Here's the stretched answer. A Mandela is energy, the budda image can be a very good explanation example of this quantifiable result of the classic Mandela image, so too are we. The universe as explained by some researches is filled with beings or entities of different varying consistencies of ether atmosphere's or energy felt and perhaps not felt. Quantifiable reasoning is varied and possibly erratic based on the scope of what is researched.

What can be attributed to the ether plain, the microcosm and macrocosm is the space bodies, just as there is an outer space, there can be an inner space with the same causation of Mandela taking place? Is the energy that has been explored in the more local areas such as science and alchemy probed this as well. A perpetual pattern of the same and not the same out of phase and in phase at the same time gradually moulding and morphing into the nature we possess all around us. An over lapping of this microcosm, macrocosm alchemy demonstrates the extreme pole of opposites in ourselves and others, from the constructs of darkness and light, all with the illusion within and without, within the space bodies and without the space bodies.

The Archon and Nature demonstrates that the whole field of the universe is geared toward entity establishment, these range from physical fixed bodies, for example a planet, a more subtle field, a nebula and to finally invisible fields of matter but may be seen through a microscope, infer-red and other spectrum analysis. The fields consist of subtle life energies that may or may not be seen, when again dealing with the macrocosm and microcosm in analysis, then the whole idea of the inner terrestrial universes found at the microscopic level of for an example, that of bacteria and smaller, that within these fields, that it may introduce a conclusive look at details dealing with the spectrum of all existences of life through the Mandela and in effect how things better function.

The Mandela has with it the same scope as the grid of reference in that it can be a determinant of distance structure and shape, thought the whole idea of it fits in with the longitude and latitude of determined points. Archon energy and vibration is much the same as ours, though it would be operating at quite possibly a higher frequency, due to the idea or the fact that it is on another plain of existence or dimension. When toying with the idea, Archon entities are represented through this continuum of a logic that is part truth and part non-truth when dealing with it on the physical plain of existence.

Reference to plains and sub-plains, which are all very prevalent on the space and grid rooms and for example; space area atmospheres, can explain the different qualities and frequencies of a message. The message that the Archons want us to hear, would be completely irrational to the normal mind. The mind is a transmitter and receiver and this determines whether we get the message or not but to the Archon energies, depending where they operate from in terms of space and time, will be doing for example work on higher or lower area vibratory rate in the aura and chakras regions of the body are all of one vibration angled in on the keying in of light or using the perspective of 0 and 1 as explained in the first chapter of my book, the base e grouping of 0 and 1 binary, there is the sound tone spectrum and frequency of the colour spectrum, within would produce it's unique energy according to the Archon receiving station outside of itself and to us as Human beings. When received the interpretation according the field will be decrypted and interpreted by the Human mind. The Archon may have a completely myopic and benevolent means of sending and packaging the message. Quantum theory has many explanations, though when Archon energy is explored we can find the answer different to what we possibly expect, expectation of a result may or may not give an answer that is workable. As with the atom explanation in illustration 11C, with the theoretical change of an atom, just as a cell in the changing of this to a human embryo, so too is the change prevalent though can be varied regarding Archon energy. So too there may be a macrocosm and microcosm at work with them. Lastly illustration 12, shows the infinite negative and positive and which as an energy, physical or non physical on whatever frequency from this illustration is what radiates out is the prolonging condition of infinity which moves out at a quantum level from any particular circular or point of radiating direction affecting and changing whatever it comes into association with and with this, the equal opposite effect with illustration 11C. This demonstrates the physical universe, the synergy of centrifugal energy. *Illustration 12.*

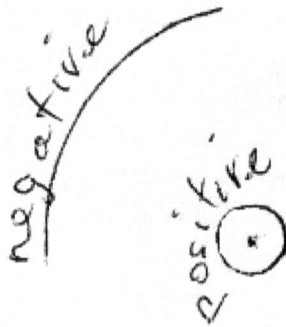

Illustration 12, the synergy of centrifugal energy.

In the 'Nag Hammadi', we have the portrayal of the Biblical character of Eve, who corresponded with the interested Archon, through a means of keeping apart from it physically, this is when it had in mind, a coupling of its energy with hers. She created another body apart from herself to couple to the Archon as the story goes and its wishes for coupling. The premise of the story is, the Archon is an energy, though the frequency is apart from the Human races and the nature of these things are found in texts going back to Qumran and the Dead sea scrolls.

Chapter 11.Conclusion

1.What result from all the information presented, is co-existent on the factor of frequency, the application, or reverberation of response to the space of any one entity or entities, whether it be an atom (female, hermaphrodite or male, organic or inorganic).

2. It's application to the algorithm in a structural and/or a morphed form, this could be an emotion to the building of an application via creation for example; sacred geometry and within that produce architecture, design, technology and mapping and language for the written form, alphabetic numerical and logarithm for the application of these in the long term sense and that of the Akashic record of knowledge of the universe. In the future as science and application of technology architecture and mapping prevail, the continuance of religious and philosophic ground will enmesh further, whether it is rough or smooth in its production this can be determined with time. Love in the correspondence as an emotion has a part to the use of application of the above mentioned to bind and make sense of it all.

3. Quantum level frequencies or tones are found throughout the twenty six letters of the alphabet to effect change at the quantum level (this can manifest in the different paradigms of existence). From one quantum whole as is binary zero's and one's, the female and the male to effect the structure as illustrated and described in illustration 11C and 12 to produce in the final conclusion the child or hermaphrodite. The three is a structure very natural to the universal principle. 3 x 3 is 9. 9 is related to the unfolding of the golden mean principle and is found in every geometric, geomantic and geographical and at an individual level, every man, woman and child is '3' and every multiplication of it i.e.; 3 x 3 can be looked at as 2 x 2 for the male and female principle is '4' double that, it's '8' for the infinite unfolding as well, symbolically. Now there is with the 3 x 3, the '9' and demonstrates individual, is microcosm and can be macrocosm though more suited to the former, the geographical, etc or 'geo' states are representative of the macrocosm, again can have instances of the microcosm. Stages cumulate to the end of a cycle and in the numbers, demonstrates this. The square or one, fuses with the circle and in the end, the principle has a 'square effect' as I call it, so for i.e.; '6', that can represent the zero and a one on inspection of that character number if we get into the detail of the symbol of that character number '6' and although has the same but opposite symbol of the '9'. When looked at on a continual basis, every number is the same in subtlety though the principle of nature found in the squaring of the circle is found mainly in the zero and the one and is as quantified as the principled frequency or tone as explained earlier in this chapter. From the quantum perspective, it's understood.

The tie in with the universe when looked at the principle of number, letter and tone/ frequency, the orbit for the twenty six tonal sounds are radiated out until critical mass changes this completely to effect the next tonal frequency. If in theory, another tonal frequency altogether, sort of comparing capacity with frequency to effect for an example,

33

another form, or change, much like what was outlined in illustration 11C to give another set of symbol energy.

The End.

~~~

Thank you for reading my book, if you want further information and updates, please write to the author at the following address; Nick Betar, c/o Australia Fair post office, Australia Fair, Qld 4215 Australia.

####

www.ingramcontent.com/pod-product-compliance
Lightning Source LLC
Chambersburg PA
CBHW060705280326
41933CB00012B/2317